S. WINSTON. POLACKS HOUSE.

7/6

AF838378

IRON
HONEY
GOLD

1

Three lives hath one life—
Iron, honey, gold.
The gold, the honey gone—
Left is the hard and cold.

ISAAC ROSENBERG

Other Works by the Compiler

POETRY

Imaginings Putnam 1961

Against the Cruel Frost Putnam 1963

FICTION

Lights in the Sky Country Putnam 1962

CRITICISM

Llareggub Revisited Bowes and Bowes 1962

The Quest for Love Methuen 1964

EDUCATION

English for Maturity Cambridge 1961

English for the Rejected Cambridge 1964

The Secret Places Methuen 1964

COMPILATIONS

Children's Games Gordon Fraser 1957

People and Diamonds Cambridge 1962

Thieves and Angels Cambridge 1962

Visions of Life Cambridge 1964

IRON HONEY GOLD

The uses of verse

AN ANTHOLOGY COMPILED BY
DAVID HOLBROOK

*

CAMBRIDGE
At the University Press
1965

PUBLISHED BY
THE SYNDICS OF THE CAMBRIDGE UNIVERSITY PRESS
Bentley House, 200 Euston Road, London, N.W. 1
American Branch: 32 East 57th Street, New York 22, N.Y.
West African Office: P.O. Box 33, Ibadan, Nigeria

PREFACE AND NOTES
©
CAMBRIDGE UNIVERSITY PRESS
1961

First published in two volumes, 1961
This edition, 1965

Cover pattern by S. M. Cockerell

Printed in Great Britain at the University Printing House, Cambridge
(Brooke Crutchley, University Printer)

For
R.B.F., R.O'M. & D.T.
and our children

CONTENTS

The first lines of the poems are given here

PREFACE

From the time we are babies we have an appetite for rhythm in words, for poetry. When we say our first word 'Mum-mum', the effect on our parent is remarkable—she laughs and claps her hands, and shouts, 'Come here! Baby has said his first words! He talks!' So we get a great satisfaction out of the simple rhythm of that repeated 'mum-mum-mum' and its effect on Mum. Of course, at first, we are making the movements of our lips as we do when we are feeding (we still say 'yum-yum') and the two great pleasures, of power through speaking, and satisfaction from food, go together. From that time forward we are devoted to poetry, to the rhythmical use of words, making patterns in words—to say things about the life around us. And not only to say things about what we see, but what we *might* see. For while we know more or less what to expect round the next corner, we can't always be sure. And we know that we can't always be sure, too. Even Dad, who seems to know everything, can't be sure that if he walks down to the bus he won't meet an escaped circus tiger.

Sometimes the world seems to be a dream, less real than the world inside us, which is all fairies and sleeping beauties. Sometimes we are told that 'facts', and something called 'science', disprove that fairies and sleeping beauties exist. Yet we are all half prepared to believe in flying saucers and men from Mars. True science has always paid great respect to the things men imagine. We should never despise imagination, for our dreams are an indispensable means to prepare ourselves for what we may meet in life. A boy who has dreamt about awakening the most wonderful princess in the world, or about throwing a bridge across the Channel, will perhaps, *because of his dreams*, become the best of husbands and the most remarkable of engineers. The girl who has dreamt of nursing a wounded lion may become, *because of her dreams*, the best mother or nurse. The scientist needs to dream above all—of a world without disease, or limitless cheap power—as well as be patient and exact. Look how long men have dreamed about going to the moon: and now we are saving up to buy our tickets, virtually.

So little Jonathan, at age three, is doing something very important when he goes round the house saying

He hi hen hand he hew-ni-horn
Har height-in hor ha hown:
Ha hi hen heat ha hew-ni-horn
Hall hound a hout he hown.

This is his version of *The Lion and the Unicorn*. While he sings what is nonsense to us he knows what a lion and a unicorn are, because he notices them on his coronation mug. He knows what fighting is, and a drum reminds him of 'drums them out of town'. His piece of rhythmical jingle helps him feel excited and powerful. He can say the verse through till it's complete, as it is when 'hown' rhymes with 'hown', and so makes a whole pattern. This pattern he can repeat to himself whenever he sees a lion, or fighting, or plum-cake, or drums. These he knows about, and has *got* them in his mind—they belong to him, because he can put them into a rhyme.

And so he feels powerful, in the face of many troubling things in life. The rhyme is a complete pattern: so the things he meets in life fall into pattern too, and he feels safer. Safer, because he knows what to expect—things come round again and again, in just the same way as the pattern of his nursery rhymes repeats itself. And children know that because someone has made up a song about something, then other people must have had the same thing happen to them as we have had happen to us: we fall down, but so did Jack and Jill and Humpty Dumpty. Later, as we grow up, we are very grateful to those people who have had experiences, joyful or sad, and put them down in verse for us. We can then share our sorrow with them: and we know that if they survived well enough to write a poem about it, then we can survive as well.

The poems in this anthology are chosen because they give us the feeling 'right through', that someone very wide awake and full of understanding has been through some happening or state of feeling in life and made verses about it for us. In Scotland they used to call poets 'makers' because they made verses out of their heads, as architects make buildings out of their heads, out of imagination. The word 'makers' is a tribute to the poets, because it surely means that the things they made were useful: in our time many people seem to think poetry is useless—but that, of course, is just their own loss. A poet knows poetry to be very useful. He makes up poems because he feels deeply not only about how he lives, but about what things everybody has to suffer, enjoy, and put up with, in life. Of course, many poets cared little about publishing their work, and others felt their poems were a tribute to God, and that was enough. But we must not be misled by this air of disparaging their own work: it was a way to avoid the worldly vanity—the conceit—which might have lured them into not being able to write any more good poetry. And the modesty goes with the sympathy for human beings; for instance, the poet Gerard Manley Hopkins wrote of his poetry to his friend Robert Bridges in 1879: 'All therefore that I think of doing is to keep my verses together in one place—at present I have not even correct copies—that, if anyone should like, they might be published after my death. And that again is unlikely, as well as remote.' George Herbert, a man who gave up being a courtier and the chance of being a famous musician to be a parish priest, wrote to a friend of his poems that 'they and I are less than the least of God's mercies', and sent from his death-bed a request that the unpublished manuscript of poems should be burnt unless 'he can think it may turn to the advantage of any dejected poor soul'.

Poets write to put their own souls in order: but because they use the language everybody speaks, they are at the same time helping put in order the soul of anyone who uses that language and chooses to read them. So in writing, poets are doing something very much like what is done by an explorer or surgeon or nurse—they are trying to help humanity, to do something 'to the advantage of any poor soul'. Even the saddest poem of all will, in 'sending' us, make us happier—because it shows us that others have lived through the most terrible experiences and have yet been able to write about them.

For instance, in this anthology is a poem by Sir Walter Raleigh, written the evening before he was going to have his head struck off by the axe. He was able to write:

> And since my flesh must die so soon
> And want a head to dine next noon,
> Just at the stroke when my veins start and spread,
> Set on my soul an everlasting head....

and he writes of how he and other Christians, after death, will take their bottles of salvation,

> And when our bottles and all we
> Are filled with immortality,
> Then blessèd paths we'll travel
> Strowed with rubies thick as gravel....

It is a marvellous and confident poem-dream: but far from being morbid and distressed in the condemned cell, he is sane enough to have wit, to joke, and to leave us a message of courage and hope that will last for ever—or while men last.

This is the power of the word, and it is seen best in the power of poetry. You will have seen messages from the past cut in old beams and walls. Here in my village a message is cut in the stone of the church tower, about the Black Death—the plague—of 1349, and a great wind of 1361: it is in Latin and I give a translation below it:

> M.Ct.Xpenta. miseranda, ferox. violenta
> M.CCCL.

Superest.plebs.pessima.testis.infineque.ventus
Validus.oc.anno.maurus.in.orbe.tonat.
M.CCCLXI

1350. Miserable, wild, distracted. 1350.
The dregs of a people alone survive to witness: and in the end a tempest
Full mighty. This year, 1361, St Maur thunders in the world.
(Graffiti in St Mary the Virgin, Ashwell.)

It speaks more history to me than any history book I have ever read. In the Keep of Norwich Castle is a message in old French painfully cut in the stone that says

Bartholomeus pur veris a tort e saunz resun
ie sui clos en cete mesun.

I Bartholomew truly by wrong and without reason
am I shut up in this prison.

Poems like Raleigh's, and messages in stone like these, and many of the songs and ballads in this anthology, speak of the sufferings and pain and death of men in the past. Of course, many others sing of love and happiness. What they all offer us is wisdom: the wisdom of the English-speaking people. The people are dead that wrote most of them; but they have left songs from their experience of life, to help us.

They have left words which tell us, in their rhythm, of the way things were—we can hear their hearts beating, their minds working, see the things they saw, in the words. Take, for instance, the last line of the first poem in volume three of this anthology:

Pale ravener of horrible meat.

The line, as you say it, makes your mouth move like a shark biting: and it makes you feel a bit sick, too (perhaps the 'v' does that)—so you feel how ghastly the shark is, because of the queer look of the creature ('pale') and the rabid way it ravenously eats any disgusting meat. Or you may hear the rattle of the train in *Casey Jones*:

That the hogger at the throttle was Casey Jones.

Or you can hear in the lines of Edward Thomas' poem *Melancholy* the way his mind was full of misery, coming and going like the noise of the wind:

> The rain and wind, the rain and wind, raved endlessly.
> On me the summer storm, and fever, and melancholy
> Wrought magic.

In these lines the rhythm is giving us in a kind of dream the feelings of the state in which the poet was under his fever. Here the lines with their rhythm give us the feelings of the rough and terrible courage of Admiral Benbow

> Brave Benbow lost his leg
>   By chain shot, by chain shot,
> Yet all the pain he bore
>   He values not...

This rough movement is like a man staggering about in terrible pain. In Wordsworth's Sonnet *Surpris'd by Joy* we can hear him turn to his little girl and then groan, as he suddenly remembers she is dead (this poem is not in this anthology: you should bear in mind that any anthology can only be a nibble at the whole feast of verse):

> Surpris'd by joy, impatient as the wind
> I turn'd to share the transport—O with whom
> But thee, deep buried in the silent tomb...

In *The Derby Ram* we can enjoy the blustering beat of lines we know, and like to be, Great Lies:

> The wool upon this sheep's back
> It grew up to the sky,
> The eagles built their nests in it,
> I heard the young ones cry.

Here, then, are the voices of men and women from the past, in all kinds of moods. They are saying to you, 'Life is like this'. And because they are poets, writing rhythmically because they are caught up in their dreams, they 'send' you, as we might say nowadays—that is, they make you feel as they did. You live through a bit of their life. So poetry doesn't say, 'When you fall in love, be careful you don't get your heart broken!', it says:

'Tis down in yonder garden green,
Love, where we used to walk:
The finest flower that ere was seen
Is withered to a stalk.

And you can take it or leave it. 'Life was like this to me: it may be to you. Best wishes, my dear.' That's what the poet says. And so say the many writers of songs who are hidden under the word 'anonymous', meaning that no-one knows the author. And what all these writers say to us in poetry is as useful as baby's first 'Mum-mum' is to him—it is a help in growing up to be adult.

1

The Pedlar of Spells

LU YU

Translated by Arthur Waley

An old man selling charms in a cranny of the town wall;
He writes out spells to bless the silkworms and spells
to protect the corn.
With the money he gets each day he only buys wine;
But he does not worry when his legs get wobbly,
for he has a boy to lean on.

2

A Widow's Weeds

WALTER DE LA MARE

A poor old Widow in her weeds
Sowed her garden with wild-flower seeds;
Not too shallow, and not too deep,
And down came April drip-drip-drip.
Up shone May, like gold, and soon
Green as an arbour grew leafy June.
And now all summer she sits and sews
Where willow-herb, comfrey, bugloss blows,
Teasle and tansy, meadowsweet,
Campion, toadflax, and rough hawksbit;
Brown bee orchis, and Peals of Bells;
Clover, burnet, and thyme she smells;
Like Oberon's meadows her garden is
Drowsy from dawn till dusk with bees.
Weeps she never, but sometimes sighs,
And peeps at her garden with bright brown eyes;
A poor old Widow in her weeds.

3

Birds at Winter Nightfall

THOMAS HARDY

Triolet

Around the house the flakes fly faster,
And all the berries now are gone
From holly and cotonea-aster
Around the house. The flakes fly!—faster
Shutting indoors that crumb-outcaster
We used to see upon the lawn
Around the house. The flakes fly faster,
And all the berries now are gone!

4

As Adam Early in the Morning

WALT WHITMAN

As Adam early in the morning,
Walking forth from the bower refresh'd with sleep,
Behold me where I pass, hear my voice, approach,
Touch me, touch the palm of your hand to my body as I pass,
Be not afraid of my body.

5

The Derby Ram

ANONYMOUS

The music is on p. 64

As I went down to Derby town,
'Twas on a market day,
And there I met the finest ram
Was ever fed on hay.
Riddle to my rye, riddle to my rye.

The wool upon this sheep's back
It grew up to the sky,
The eagles built their nests in it,
I heard the young ones cry.

The horns upon this ram's head
They grew up to the moon;
The preacher went up in April
And never was down till June.

The wool upon this sheep's tail
Was very fine and thin;
It took all the ladies in Derby town
Seven years to spin.

This ram he had four feet
And on them he did stand;
And every foot he had
Would cover an acre of land.

Every tooth this ram had
Was hollow as a horn;
They took one out and measured it,
It held a barrel of corn.

This ram's head was
As big as Noah's ark;
My dog ran up his nostrils,
And turned around and barked.

The one that killed this ram
Was up to his knees in blood;
And the one that held the basin
Was washed away in the flood

It took all the men in Derby town
To haul away his horns;
And all the ladies in Derby town
To roll away his bones.

And if you don't believe me
And think it is a lie;
You go down to Derby town
And see as well as I.

6

As ye Came from the Holy Land

SIR WALTER RALEIGH

As ye came from the holy land
  Of Walsinghame,
Met you not with my true love
  By the way as you came?

How should I know your true love,
  That have met many a one,
As I came from the holy land,
  That have come, that have gone?

She is neither white nor brown,
  But as the heavens fair:
There is none hath her form divine
  In the earth or the air.

Such a one did I meet, good sir,
  Such an angelic face,
Who like a nymph, like a queen, did appear
  In her gait, in her grace.

She hath left me here alone
   All alone, as unknown,
Who sometime did me lead with herself,
   And me loved as her own.

What's the cause that she leaves you alone
   And a new way doth take,
That sometime did love you as her own,
   And her joy did you make?

I have loved her all my youth,
   But now am old, as you see:
Love likes not the falling fruit,
   Nor the witherèd tree.

Know that Love is a careless child,
   And forgets promise past:
He is blind, he is deaf when he list,
   And in faith never fast.

His desire is a dureless content,
   And a trustless joy;
He is won with a world of despair,
   And is lost with a toy.

Of womankind such indeed is the love,
   Or the word love abusèd,
Under which many childish desires
   And conceits are excusèd.

But true love is a durable fire,
   In the mind ever burning,
Never sick, never dead, never cold,
   From itself never turning.

7

The Shad

ANONYMOUS

The music is on p. 65

Bait a hook to catch a shad,
The first thing he bit was my old Dad.
Pulled her away with all my might,
Trying for to get the old man out.
Fish-pole broke and I got mad,
Down to the bottom went old Dad.

8

Thunder

WALTER DE LA MARE

Call the cows home!
Call the cows home!
Louring storm clouds
Hitherward come;
East to West
Their wings are spread;
Lost in the blue
Is each heaven-high head;
They've dimmed the sun;
Turned day to night;
With a whistling wind
The woods are white;
Down streams the rain
On farm, barn, byre,
Bright green hill,
And bramble and briar,

Filling the valley
With glimmer and gloom:
Call the cows home!
Call the cows home!

9

To Mrs Reynolds's Cat

JOHN KEATS

Cat! who hast pass'd thy grand climacteric,
  How many mice and rats hast in thy days
  Destroy'd?—How many tit bits stolen? Gaze
With those bright languid segments green, and prick
Those velvet ears—but pr'ythee do not stick
  Thy latent talons in me—and upraise
  Thy gentle mew—and tell me all thy frays
Of fish and mice, and rats and tender chick.
Nay, look not down, nor lick thy dainty wrists—
  For all the wheezy asthma,—and for all
Thy tail's tip is nick'd off—and though the fists
  Of many a maid have given thee many a maul,
Still is that fur as soft as when the lists
  In youth thou enter'dst on glass bottled wall.

10

Up in the Morning Early

ROBERT BURNS

Cauld blaws the wind frae east to west,
  The drift is driving sairly;
Sae loud and shrill's I hear the blast—
  I'm sure it's winter fairly.

Up in the morning's no for me,
  Up in the morning early;
When a' the hills are covered wi' snaw,
  I'm sure it's winter fairly.

The birds sit chittering in the thorn,
  A' day they fare but sparely;
And lang's the night frae e'en to morn—
  I'm sure it's winter fairly.

Up in the morning's no for me,
  Up in the morning early;
When a' the hills are covered wi' snaw,
  I'm sure it's winter fairly.

II

Winter Warfare

EDGELL RICKWORD

Colonel Cold strode up the Line
  (tabs of rime and spurs of ice);
stiffened all who met his glare:
  horses, men, and lice.

Visited a forward post,
  left them burning, ear to foot;
fingers stuck to biting steel,
  toes to frozen boot.

Stalked on into No Man's Land,
  turned the wire to fleecy wool,
iron stakes to sugar sticks
  snapping at a pull.

Those who watched with hoary eyes
   saw two figures gleaming there;
Hauptmann Kalte, Colonel Cold,
   gaunt in the grey air.

Stiffly, tinkling spurs they moved,
   glassy eyed, with glinting heel
stabbing those who lingered there
   torn by screaming steel.

12

Christmas Eve: A Ceremonie

ROBERT HERRICK

Come guard this night the Christmas-pie,
That the Thiefe, though ne'r so slie,
With his flesh-hooks, don't come nigh
        To catch it.

From him who all alone sits there,
Having his eyes still in his eare,
And a deal of nightly fear
        To watch it.

13

Dance to your Daddy

ANONYMOUS

Dance to your daddy,
My little laddy.
Dance to your daddy,
My little man.
Thou shalt have a fish,
Thou shalt have a fin,
Thou shalt have a haddock
When the boat comes in.

Thou shalt have a codling,
Boilèd in a pan,
Dance to your daddy
My little man.
When thou art a man,
And fit to take a wife,
Thou shalt wed a maid
And love her all your life.
She shall be your lassie—
Thou shalt be her man.
Dance to your daddy
My little man.

14

Dearly Beloved Brethren

JOSEPH TABRAR

Dearly beloved brethren isn't it a sin
When you peel potatoes you throw away the skin,
The skins feed pigs and the pigs feed you:
Dearly beloved brethren is not that true?

15

Down by the Salley Gardens

W. B. YEATS

Down by the salley gardens my love and I did meet;
She passed the salley gardens with little snow-white feet.
She bid me take love easy, as the leaves grow on the tree;
But I, being young and foolish, with her would not agree.

In a field by the river my love and I did stand,
And on my leaning shoulder she laid her snow-white hand.
She bid me take life easy, as the grass grows on the weirs;
But I was young and foolish, and now am full of tears.

16

Tarra Ding Ding Ding Dido

ANONYMOUS

Down to the knees in blood, up to the knees in water,
  My boots are lined with gold, my stockings lined with silver;

A red rose on my breast, a gold ring on my finger,
  Tarra ding ding ding, tarra ding ding ding dido.

Down to the knees in blood, up to the knees in water,
  My boots are lined with gold, my stockings lined with silver;

I for the pots and pans, I for the man that made them,
  Tarra ding ding ding, tarra ding ding ding dido.

17

Shepherds' Counting Rhyme

ANONYMOUS

Eeny, teeny, ether, fether, fip,
  Satha, latha, ko, darthur, dick,
Ten-dick, teen-dick, ether-dick, fether-dick, bunkin,
  Een-bunkin, teen-bunkin, ether-bunkin, fether-bunkin, digit.

18

Winter Rain

CHRISTINA ROSSETTI

Every valley drinks,
  Every dell and hollow,
Where the kind rain sinks and sinks,
  Green of Spring will follow.

Yet a lapse of weeks—
  Buds will burst their edges,
Strip their wool-coats, glue-coats, streaks,
  In the woods and hedges;

Weave a bower of love
    For birds to meet each other,
Weave a canopy above
    Nest and egg and mother.

But for fattening rain
    We should have no flowers,
Never a bud or leaf again
    But for soaking showers....

19

Game Rhyme

ANONYMOUS

Green peas, mutton pies, tell me where my Maggie lies,
    I'll be there before she dies, green peas, mutton pies:
Three pairs of blankets, and four pairs of sheets,
    One yard of cotton to mend my Johnny's breeks.

Green peas, mutton pies, tell me where my Johnny lies,
    I'll be there before he dies and cuddle in his bosom.
Baby in the cradle, playing with the keys,
    Maggie in the pea park, picking up the peas.

20

The Eagle

ALFRED LORD TENNYSON

He clasps the crag with hookèd hands
Close to the sun in lonely lands;
Ring'd with the azure world he stands.
The wrinkled sea beneath him crawls;
He watches from his mountain walls,
And like a thunderbolt he falls.

21

Song from 'Charles the First'

P. B. SHELLEY

Heigho! the lark and the owl!
  One flies the morning, and one lulls the night:—
Only the nightingale, poor fond soul,
  Sings like the fool through darkness and light.

'A widow bird sate mourning for her love
  Upon a wintry bough;
The frozen wind crept on above,
  The freezing stream below.

'There was no leaf upon the forest bare,
  No flower upon the ground,
And little motion in the air
  Except the mill-wheel's sound.'

22

The House of Hospitalities

THOMAS HARDY

Here we broached the Christmas barrel,
  Pushed up the charred log-ends;
Here we sang the Christmas carol,
  And called in friends.

Time has tired me since we met here
  When the folk now dead were young
Since the viands were outset here
  And quaint songs sung.

And the worm has bored the viol
  That used to lead the tune,
Rust eaten out the dial
  That struck night's noon.

Now no Christmas brings in neighbours,
And the New Year comes unlit;
Where we sang the mole now labours,
And spiders knit.

Yet at midnight if here walking,
When the moon sheets wall and tree,
I see forms of old time talking,
Who smile on me.

23

Mary Ann

JOSEPH TABRAR

He's bought a bed and a table too,
A big tin dish for making stew,
A large flat-iron to iron his shirt,
And a flannel, and a scrubbing brush to wash away the dirt.
And he's bought a pail and basins three,
A coffee pot, a kettle, and a teapot for the tea,
And a soap-bowl and a ladle,
And a gridiron and a cradle,
And he's going to marry Mary Ann, that's me!
He's going to marry Mary Ann!

24

Counting out Rhyme

ANONYMOUS

Horcum borcum curious corkum
Herricum, berricam, buzz:
Eggs, butter, cheese, bread
Stick, stock, stone dead.

25

The Queen of Hearts

CHRISTINA ROSSETTI

How comes it, Flora, that, whenever we
Play cards together, you invariably,
  However the pack parts,
  Still hold the Queen of Hearts?

I've scanned you with a scrutinising gaze,
Resolved to fathom these your secret ways:
  But, sift them as I will,
  Your ways are secret still.

I cut and shuffle; shuffle, cut, again;
But all my cutting, shuffling, proves in vain:
  Vain hope, vain forethought too;
  That Queen still falls to you.

I dropped her once, prepense; but, ere the deal
Was dealt, your instinct seemed her loss to feel:
  'There should be one card more,'
  You said, and searched the floor.

I cheated once; I made a private notch
In Heart-Queen's back, and kept a lynx-eyed watch;
  Yet such another back
  Deceived me in the pack:

The Queen of Clubs assumed by arts unknown
An imitative dint that seemed my own;
  This notch, not of my doing,
  Misled me to my ruin.

It baffles me to puzzle out the clue,
Which must be skill, or craft, or luck in you:
  Unless, indeed, it be
  Natural affinity.

26

Ana-{MARY / ARMY}gram

GEORGE HERBERT

How well her name an *Army* doth present,
In whom the *Lord of hosts* did pitch his tent!

27

If all the World were Paper

ANONYMOUS

If all the world were paper
  If all the sea were ink,
If all the trees were bread and cheese,
  What would we have to drink?

If all the vessels ran,
  If none but had a crack,
If Spanish apes ate all the grapes,
  What would we do for sack?

28

If it's ever spring again

THOMAS HARDY

If it's ever spring again,
    Spring again,
I shall go where went I when
Down the moor-cock splashed, and hen,
Seeing me not, amid their flounder,
Standing with my arm around her;
If it's ever spring again,
    Spring again.
I shall go where went I then.

If it's ever summer-time,
    Summer-time,
With the hay crop at the prime,
And the cuckoos—two—in rhyme,
As they used to be, or seemed to,
We shall do as long we've dreamed to,
If it's ever summer-time,
    Summer-time,
With the hay, and bees achime.

29

If the Man who Turnip Cries

SAMUEL JOHNSON

If the man who turnip cries
Cry not when his father dies
'Tis a proof that he had rather
Have a turnip than a father.

30

The Song of the Old Mother

W. B. YEATS

I rise in the dawn, and I kneel and blow
Till the seed of the fire flicker and glow;
And then I must scrub and bake and sweep
Till stars are beginning to blink and peep;
And the young lie long and dream in their bed
Of the matching of ribbons for bosom and head,
And their day goes over in idleness,
And they sigh if the wind but lift a tress:
While I must work because I am old,
And the seed of the fire gets feeble and cold.

31

Cushie Butterfield

GEORGE RIDLEY

I's a broken-hearted keelman, and I's over head in love
With a young lass in Gateshead and I call her my dove.
Her name's Cushie Butterfield, and she sells yellow clay,
And her cousin is a muckman, and they call him Tom Grey.

She's a big lass, and a bonny lass,
And she likes her beer,
And they call her Cushie Butterfield
And I wish she was here.

Her eyes is like two holes in a blanket burnt through,
Her brows in a morning would spyen a young cow,
And when I hear her shouting, 'Will ye buy any clay?'
Like a candyman's trumpet, it steals my heart away.

Ye'll oft see her down at Sandgate when the fresh herring come in.
She's like a bag full of sawdust, tied round with a string.
She wears big galoshes too, and her stockings once was white,
And her petticoat's lilac, and her hat's never straight.

When I asked her to marry me she started to laugh;
'No! None of your monkey tricks, for I don't like such chaff.'
Then she starts a-blubbering, and she roared like a bull,
And the chaps on the Quay says I'm nowt but a fool.

She says the chap that gets her must work every day,
And when he comes home at nights he must go and seek clay;
And when he's away seeking, she'll make balls and sing
'Oh well may the keel row that my laddie's in.'

32

Miss T.

WALTER DE LA MARE

It's a very odd thing—
  As odd as can be—
That whatever Miss T. eats
  Turns into Miss T.;
Porridge and apples,
  Mince, muffins and mutton,
Jam, junket, jumbles—
  Not a rap, not a button
It matters; the moment
  They're out of her plate,
Though shared by Miss Butcher
  And sour Mr. Bate;
Tiny and cheerful,
  And neat as can be,
Whatever Miss T. eats
  Turns into Miss T.

33

A Little Boy Threw

ANONYMOUS

It snowed a mist, it snowed a mist,
It snowed all over the land:
Till all the boys throughout the town
Went out with ball in hand.

A little boy threw his ball so high,
He threw his ball so low;
He threw it into a dusky garden
Among the blades of snow.

'Come here, come here, you sweet little boy,
Come here and get your ball.'
'I'll not come here and I'll not come there,
And I'll not come in your hall.'

She showed him an apple as yellow as gold,
She showed him a bright gold ring,
She showed him a cherry as red as blood,
And that enticed him in;

Enticed him into the sitting room,
Enticed him into the kitchen,
And there he saw his own dear nurse
A-picking of a chicken.

'I'm washing this basin the livelong day,
To catch your heart's blood in;
And I won't spare you nor yet your ma,
Nor any of your kin.'

And she dragged him to the cooling board,
And stabbed him like a sheep,
And threw him into the dusky well,
With his grammar at his feet.

34

The Sandgate Girl's Lamentation

ANONYMOUS

The music is on p. 66

I was a young maid truly,
And lived in Sandgate Street.
I thought to marry a good man
To keep me warm at neet.

He's an ugly body, a bubbly body,
An ill-fared, hideous loon;
And I have married a keelman,
And my good days are done.

I thought to marry a parson
To hear me say my prayers;
But I have married a keelman
And he kicks me down the stairs.

I thought to marry a dyer
To dye my apron blue;
But I have married a keelman
And he makes me sorely rue.

I thought to marry a joiner
To make me chair and stool;
But I have married a keelman
And he's a perfect fool.

I thought to marry a sailor
To bring me sugar and tea;
But I have married a keelman
And that he lets me see.

35

Poor Old Horse

ANONYMOUS

My clothing was once of the linsey woolsey fine,
My tail it grew at length, my coat did likewise shine;
But now I'm growing old; my beauty does decay,
My master frowns upon me; one day I heard him say,
Poor old horse: let him die.

Once I was kept in the stable snug and warm,
To keep my tender limbs from any cold or harm;
But now, in open fields, I am forced to go,
In all sorts of weather, let it be hail, rain, freeze, or snow.
Poor old horse: let him die.

Once I was fed on the very best corn and hay
That ever grew in yon fields, or in yon meadows gay;
But now there's no such doing can I find at all,
I'm glad to pick the green sprouts that grow behind yon wall.
Poor old horse: let him die.

'You are old, you are cold, you are deaf, dull, dumb and slow,
You are not fit for anything, or in my team to draw.
You have eaten all my hay, you have spoiled all my straw,
So hang him, whip, stick him, to the huntsman let him go.'
Poor old horse: let him die.

My hide unto the tanners then I would freely give,
My body to the hound dogs, I would rather die than live,
Likewise my poor old bones that have carried you many a mile,
Over hedges, ditches, brooks, bridges, likewise gates and stiles.
Poor old horse: let him die.

36

The Little Black Boy

WILLIAM BLAKE

My mother bore me in the southern wild,
And I am black, but O! my soul is white;
White as an angel is the English child,
But I am black, as if bereav'd of light.

My mother taught me underneath a tree,
And sitting down before the heat of day,
She took me on her lap and kissed me,
And pointing to the east, began to say:

'Look on the rising sun: there God does live,
And gives his light, and gives his heat away;
And flowers and trees and beasts and men receive
Comfort in morning, joy in the noonday.

'And we are put on earth a little space,
That we may learn to bear the beams of love;
And these black bodies and this sunburnt face
Is but a cloud, and like a shady grove.

'For when our souls have learn'd the heat to bear,
The cloud will vanish; we shall hear his voice,
Saying: "Come out from the grove, my love & care,
And round my golden tent like lambs rejoice."'

Thus did my mother say, and kissed me;
And thus I say to little English boy:
When I from black and he from white cloud free,
And round the tent of God like lambs we joy,

I'll shade him from the heat, till he can bear
To lean in joy upon our father's knee;
And then I'll stand and stroke his silver hair,
And be like him, and he will then love me.

37

The Flying Cloud

ANONYMOUS

My name is Edward Holland
As you may understand
I belong to the country of Waterford
Near Eirings happy land

When I was young and in my prime
And fortune on me smiled
My parents reared me tenderly
I being their only child

My father bound me to a trade
In Waterford's own town
He bound me to a cooper there
By the name of William Brown

I served my master faithfully
For eighteen months or more
Till I shipped aboard the Ocean Queen
Bound down for Bermuda's shore

When we arrived at Bermuda shore
I fell in with a Capting More,
The commander of the Flying Cloud
And belonging to Trymore.

He questioned and cross questioned me
On a slaving voyage to go
To the burning shores of Africa
Where the sugar cane does grow.

Now three or four weeks after this
We arrived at the African shore
And eighteen hundred of those poor souls
From their native land we bore.

We would march them on our quarterdecks
And store them down below.
It was eighteen inches to a man;
It was what they had to go.

We sailed away without delay
With our cargo of slaves.
It would have been better for them poor souls
Had they been in their graves.

The plaguey fever came aboard
Swept half of them away
We dragged their dead bodies on deck
And threw them in the sea.

Now two or three weeks after this
We arrived at Bermuda shore
Where we sold them to the planters there
To be slaves for evermore.

The rye and coffee fields to hoe
Beneath the burning sun
For to live a long and dreary life
'Till their career was run.

Now when our slaving money was all gone
We put to sea again.
When Capting More he came on deck
And said to us his men,

'There is gold and silver to be had
If you will agree with me,
We will hoist aloft the pirate flag
And scour the raging sea.'

Now they all agreed excepting five
And those we had to land.
Two of them being Boston boys,
Two more from Newfoundland.

The other being an Irish man
Belonging to Trymore.
I wish to God I had joined those five
And went with them on shore.

Now we robbed and plundered many the ship
Down on the Spanish Main
Caused many the widow and orphan child
In sorrow to remain.

We would march them on the quarterdecks
Give them a watery grave
For the saying of that Capting More
That a dead man tells no tales.

Now chased were we by many a ship
Both frigates and liners too.
It was all in vain astern of us
Their bomb shells they did throw.

It was all in vain astern of us
Their cannons roared aloud.
It was all in vain for to ever try
For to catch the Flying Cloud.

Now the Flying Cloud was a Spanish ship
Five thousand tons or more.
She would easily outsail any ship
Coming out of Baltimore.

Her sails were as white as the driven snow
On them there were no stains.
With a forty-nine brass pounder gun
She would carry an after main.

Now the Flying Cloud was as fine a ship
As ever swam the sea.
Or ever spread a main top sail
Before a pleasant breeze.

Now often have I watched that ship
As she went sailing by
With her royal galyants spread aloft
Above the canvas high.

Now a Spanish ship, the man of war
Her dungeon hove in view.
She fired a shot acrost our deck
As a signal to heave to.

But to her we paid no attention
As before a pleasant breeze
Till a chance shot cut our mainmast down
We then soon fell behind.

We cleared our decks for action
As she ranged up alongside.
And soon acrost our quarterdecks
There flowed a crimson tide.

We fought till Capting More was slain
And eight of our men
Till a bombshell set our ship on fire
We were forced to surrender then.

It's now to Newgate we are brought
Bound down in iron bands
For the murdering and plundering of many the ship
Down on the ocean strand.

It's drinking and bad company
That's made a wretch of me.
Come all young men a warning take
Bid a curse to the pirate sea.

38

Lament of Hsi-Chun

Translated by Arthur Waley

About the year 105 B.C. *a Chinese lady named Hsi-Chun was sent to be the wife of a central Asian nomad king, K'un Mo, king of the Wu-sun. When she got there, she found her husband was old and decrepit. He only saw her once or twice a year when they drank a cup of wine together. They could not talk as they had no language in common.*

My people have married me
In a far corner of Earth;
Sent me away to a strange land,
To the king of the Wu-sun.
A tent is my house,
Of felt are my walls;
Raw flesh my food
With mare's milk to drink.
Always thinking of my own country,
My heart sad within.
Would I were a yellow stork
And could fly to my old home!

39

Fine Work with Pitch and Copper

WILLIAM CARLOS WILLIAMS

Now they are resting
in the fleckless light
separately in unison

like the sacks
of sifted stone stacked
regularly by twos

about the flat roof
ready after lunch
to be opened and strewn

The copper in eight
foot strips has been
beaten lengthwise

down the center at right
angles and lies ready
to edge the coping

One still chewing
picks up a copper strip
and runs eye along it

40

Oats and Beans and Barley

ANONYMOUS

Oats and beans and barley grow!
Oats and beans and barley grow!
Do you or I or anyone know
How oats and beans and barley grow?

First the farmer sows his seed,
Then he stands and takes his ease,
Stamps his foot and *claps* his hands,
And *turns around* to view the land.
Waiting for a partner,
Waiting for a partner!
Open the ring and take one in!

Now you are married you must obey,
You must be true to all you say.
You must be kind, you must be good
And help your wife to chop the wood!

41

Lovewell's Fight

ANONYMOUS

Of worthy Captain Lovewell, I purpose now to sing,
How valiantly he served his country and his King;
He and his valiant soldiers, did range the woods full wide,
And hardships they endured to quell the Indians' pride.

'Twas nigh unto Pigwacket, on the eighth day of May,
They spied a rebel Indian soon after break of day;
He on a bank was walking, upon a neck of land,
Which leads into a pond as we're made to understand.

Our men resolv'd to have him and travell'd two miles round,
Until they met the Indian, who boldly stood his ground;
Then spake up Captain Lovewell, 'Take you good heed,' says he,
'This rogue is to decoy us, I very plainly see.'

'The Indians lie in ambush, in some place nigh at hand,
In order to surround us upon this neck of land;
Therefore we'll march in order, and each man leave his pack,
That we may briskly fight them when they make their attack.'

They came unto this Indian, who did them this defy,
As soon as they came nigh him, two guns did he let fly,
Which wounded Captain Lovewell, and likewise one man more,
But when this rogue was running, they laid him in his gore.

Then having scalp'd the Indian, they went back to the spot,
Where they had laid their packs down, but there they found them not,
For the Indians having spy'd them, when they them down did lay,
Did seize them for their plunder, and carry them away.

These rebels lay in ambush, this very place hard by,
So that an English soldier did one of them espy,
And cried out 'Here's an Indian,' with that they started out,
As fiercely as old lions, and hideously did shout.

With that our valiant English, all gave a loud huzza,
To shew the rebel Indians they fear'd them not a straw;
So now the fight began, and as fiercely as could be,
The Indians ran up to them, but soon were forc'd to flee.

Then spake up Captain Lovewell, when first the fight began,
'Fight on my valiant heroes! You see they fall like rain.'
For as we are inform'd, the Indians were so thick,
A man could scarcely fire a gun and not some of them hit.

Then did the rebels try their best our soldiers to surround,
But they could not accomplish it, because there was a pond,
To which our men retreated and covered all the rear,
The rogues were forc'd to flee them, altho' they skulk'd for fear.

Two logs there were behind them, that close together lay,
Without being discovered, they could not get away,
Therefore our valiant English, they travell'd in a row,
And at a handsome distance as they were wont to go.

’Twas 10 o’clock in the morning, when first the fight begun,
And fiercely did continue until the setting sun,
Excepting that the Indians, some hours before ’twas night,
Drew off into the bushes and ceased awhile to fight.

But soon again returned, in fierce and furious mood,
Shouting as in the morning, but not yet half so loud;
For as we are informed, so thick and fast they fell,
Scarce twenty of their number at night did get home well.

And that our valiant English, till midnight there did stay,
To see whether the rebels would have another fray;
But they no more returning, they made off towards their home,
And brought away their wounded as far as they could come.

Of all our valiant English, there were but thirty-four,
And of the rebel Indians, there were about four score.
And sixteen of our English did safely home return,
The rest were killed and wounded, for which we all must mourn.

Our worthy Captain Lovewell among them there did die,
They killed Lt. Robins, and wounded good young Frye,
Who was our English chaplain; he many Indians slew,
And some of them he scalp’d when bullets round him flew.

Young Fullam too I’ll mention, because he fought so well,
Endeavouring to save a man, a sacrifice he fell;
But yet our valiant Englishmen in fight were ne’er dismay’d,
But still they kept their motion, and Wyman’s Captain made.

Who shot the old chief Paugus, which did the foe defeat,
Then set his men in order, and brought off the retreat;
And braving many dangers and hardships in the way,
They safe arriv’d at Dunstable, the thirteenth day of May.

42

Goody Blake and Harry Gill

WILLIAM WORDSWORTH

Oh! what's the matter? what's the matter?
What is't that ails young Harry Gill?
That evermore his teeth they chatter,
Chatter, chatter, chatter still!
Of waistcoats Harry has no lack,
Good duffle grey, and flannel fine;
He has a blanket on his back,
And coats enough to smother nine.

In March, December, and in July,
'Tis all the same with Harry Gill;
The neighbours tell, and tell you truly,
His teeth they chatter, chatter still.
At night, at morning, and at noon,
'Tis all the same with Harry Gill;
Beneath the sun, beneath the moon,
His teeth they chatter, chatter still!

Young Harry was a lusty drover,
And who so stout of limb as he?
His cheeks were red as ruddy clover:
His voice was like the voice of three.
Old Goody Blake was old and poor;
Ill fed she was, and thinly clad;
And any man who passed her door
Might see how poor a hut she had.

All day she spun in her poor dwelling:
And then her three hours' work at night,
Alas! 'twas hardly worth the telling,
It would not pay for candle-light.

Remote from sheltered village green,
On a hill's northern side she dwelt,
Where from sea-blasts the hawthorns lean
And hoary dews are slow to melt.

By the same fire to boil their pottage,
Two poor old Dames, as I have known,
Will often live in one small cottage;
But she, poor Woman! housed alone.
'Twas well enough when summer came,
The long, warm, lightsome summer-day,
Then at her door the *canty* Dame
Would sit, as any linnet gay.

But when the ice our streams did fetter,
Oh then how her old bones would shake!
You would have said, if you had met her,
'Twas a hard time for Goody Blake.
Her evenings then were dull and dead:
Sad case it was, as you may think,
For very cold to go to bed;
And then for cold not sleep a wink.

O joy for her! whene'er in winter
The winds at night had made a rout;
And scattered many a lusty splinter
And many a rotten bough about.
Yet never had she, well or sick,
As every man who knew her says,
A pile beforehand, turf or stick,
Enough to warm her for three days.

Now, when the frost was past enduring,
And made her poor old bones to ache,
Could any thing be more alluring
Than an old hedge to Goody Blake?

And, now and then, it must be said,
When her old bones were cold and chill,
She left her fire, or left her bed,
To seek the hedge of Harry Gill.

Now Harry he had long suspected
This trespass of old Goody Blake;
And vowed that she should be detected—
That he on her would vengeance take.
And oft from his warm fire he'd go,
And to the fields his road would take;
And there, at night, in frost and snow,
He watched to seize old Goody Blake.

And once, behind a rick of barley,
Thus looking out did Harry stand:
The moon was full and shining clearly,
And crisp with frost the stubble land.
—He hears a noise—he's all awake—
Again?—on tip-toe down the hill
He softly creeps—'tis Goody Blake.
She's at the hedge of Harry Gill!

Right glad was he when he beheld her:
Stick after stick did Goody pull:
He stood behind a bush of elder,
Till she had filled her apron full.
When with her load she turned about,
The by-way back again to take;
He started forward, with a shout,
And sprang upon poor Goody Blake.

And fiercely by the arm he took her,
And by the arm he held her fast,
And fiercely by the arm he shook her,
And cried, 'I've caught you then at last!'

Then Goody, who had nothing said,
Her bundle from her lap let fall;
And, kneeling on the sticks she prayed,
To God that is the judge of all.

She prayed, her withered hand uprearing,
While Harry held her by the arm—
'God! who art never out of hearing,
O may he never more be warm!'
The cold, cold moon above her head,
Thus on her knees did Goody pray;
Young Harry heard what she had said:
And icy cold he turned away.

He went complaining all the morrow
That he was cold and very chill:
His face was gloom, his heart was sorrow,
Alas! that day for Harry Gill!
That day he wore a riding-coat,
But not a whit the warmer he:
Another was on Thursday brought,
And ere the Sabbath he had three.

'Twas all in vain, a useless matter,
And blankets were about him pinned;
Yet still his jaws and teeth they clatter
Like a loose casement in the wind.
And Harry's flesh it fell away;
And all who see him say, 'tis plain,
That, live as long as live he may,
He never will be warm again.

No word to any man he utters,
A-bed or up, to young or old;
But ever to himself he mutters,
'Poor Harry Gill is very cold.'

A-bed or up, by night or day;
His teeth they chatter, chatter, still.
Now think, ye farmers all, I pray,
Of Goody Blake and Harry Gill!

43
Tillie

WALTER DE LA MARE

Old Tillie Turveycombe
Sat to sew,
Just where a patch of fern did grow;
There, as she yawned,
And yawn wide did she,
Floated some seed
Down her gull-e-t;
And look you once,
And look you twice,
Poor old Tillie
Was gone in a trice.
But oh, when the wind
Do a-moaning come,
'Tis poor old Tillie
Sick for home;
And oh, when a voice
In the mist do sigh,
Old Tillie Turveycombe's
Floating by.

44

The Big Rock Candy Mountains

ANONYMOUS

One evenin' as the sun went down
And the jungle fire was burnin',
Down the track came a hobo hikin',
And he said: 'Boys, I'm not turnin',
I'm headed fer a land that's far away
Beside the crystal fountains,
So come with me, we'll all go see
The Big Rock Candy Mountains.'

In the Big Rock Candy Mountains,
There's a land that's fair and bright,
Where the handouts grow on bushes,
And you sleep out every night.
Where the boxcars are all empty,
And the sun shines every day
On the birds and the bees and the cigarette trees,
And the lemonade springs where the bluebird sings,
In the Big Rock Candy Mountains.

In the Big Rock Candy Mountains,
All the cops have wooden legs,
The bulldogs all have rubber teeth,
And the hens lay soft-boiled eggs.
The farmers' trees are full of fruit,
And the barns are full of hay.
Oh, I'm bound to go where there ain't no snow,
Where the rain don't pour, the wind don't blow,
In the Big Rock Candy Mountains.

In the Big Rock Candy Mountains,
You never change your socks,
And the little streams of alcohol
Come tricklin' down the rocks.

There the brakemen have to tip their hats
And the railroad bulls are blind.
There's a lake of stew and of whisky too,
You can paddle all around 'em in a big canoe,
In the Big Rock Candy Mountains.

In the Big Rock Candy Mountains,
All the jails are made of tin,
And you can bust right out again
As soon as you are in.
There ain't no short-handled shovels,
No axes, saws or picks.
I'm goin' to stay where you sleep all day,
Where they hung the Turk that invented work,
In the Big Rock Candy Mountains.

45

One Two Three

ANONYMOUS

One two three
  Father caught a flea:
Put it in the teapot
  To make a cup of tea.

46

Thaw

EDWARD THOMAS

Over the land freckled with snow half-thawed
The speculating rooks at their nests cawed
And saw from elm-tops, delicate as flower of grass,
What we below could not see, Winter pass.

47

Frying Pan's Theology

A. B. PATERSON

SCENE: On Monaro.
  Dramatis Personae:
Shock-headed blackfellow,
  Boy (on a pony).

Snowflakes are falling
  Gentle and slow,
Youngster says, 'Frying Pan
  What makes it snow?'

Frying Pan, confident,
  Makes the reply—
'Shake 'im big flour bag
  Up in the sky!'

'What! when there's miles of it?
  Surely that's brag.
Who is there strong enough
  Shake such a bag?'

'What parson tellin' you,
  Ole Mister Dodd,
Tell you in Sunday-school?
  Big pfeller God!

'Him drive 'im bullock dray,
  Then thunder go;
Him shake 'im flour bag—
  Tumble down snow!'

48

Lady Maisry

ANONYMOUS

She called to her little page-boy,
Who was her brother's son;
She told him as quick as he could go,
To bring her lord safe home.

Now the very first mile he would walk
And the second he would run,
And when he came to a broken broken bridge
He bent his breast and swum.

Now when he came to the New Castel
The lord was set at meat
'If you were to know as much as I
How little would you eat.'

'O is my tower falling down?
Or does my bower burn?
Or is my lady lighter yet
Of a daughter or a son?'

'O no your tower is not falling down,
Nor does your bower burn:
But we are afraid ere you return
Your lady will be dead and gone.'

'Come servant saddle my milk-white steed!
Come saddle my pony too!
That I may neither eat nor drink
Till I come to the Old Castel!'

Now when he came to the Old Castel
He heard a big bell toll,
And then he saw his noble noblemen
A-bearing of a pall.

'Lay down, lay down that gentle gentle corpse'
(As it lay fast asleep)
'That I may kiss her red ruby lips
Which I used to kiss so sweet!'

Six times he kissed her red ruby lips,
Nine times he kissed her chin,
Ten times he kissed her snowy snowy breast,
Where love did enter in.

The lady was buried on that Sunday
Before the prayer begun:
And the lord he died on the next Sunday
Before the prayer was done.

49

Hailstorm in May

GERARD MANLEY HOPKINS

Strike, churl; hurl, cheerless wind, then; heltering hail
May's beauty massacre and wispèd wild clouds grow
Out on the giant air; tell Summer No,
Bid joy back, have at the harvest, keep Hope pale

50

Vertue

GEORGE HERBERT

Sweet day, so cool, so calm, so bright,
The bridall of the earth and skie:
The dew shall weep thy fall to night;
                    For thou must die.

Sweet rose, whose hue angrie and brave
Bids the rash gazer wipe his eye:
Thy root is ever in its grave,
And thou must die.

Sweet spring, full of sweet dayes and roses,
A box where sweets compacted lie;
My musick shows ye have your closes,
And all must die.

Onely a sweet and vertuous soul,
Like season'd timber, never gives;
But though the whole world turn to coal,
Then chiefly lives.

51

Tall Nettles

EDWARD THOMAS

Tall nettles cover up, as they have done
These many springs, the rusty harrow, the plough
Long worn out, and the roller made of stone:
Only the elm butt tops the nettles now.

This corner of the farmyard I like most:
As well as any bloom upon a flower
I like the dust on the nettles, never lost
Except to prove the sweetness of a shower.

52

The Crows Kept Flyin'

ANONYMOUS

The crows kept flyin' up, boys,
The crows kept flyin' up.
The dog he seen and whimpered, boys,
Though he was but a pup.

The lost was found, we brought him round
And took him from the place,
While the ants was swarming on the ground,
And the crows was sayin' grace.

53

The 23 Psalme

GEORGE HERBERT

The God of love my shepherd is,
    And he that doth me feed:
While he is mine, and I am his,
    What can I want or need?

He leads me to the tender grasse,
    Where I both feed and rest;
Then to the streams that gently passe:
    In both I have the best.

Or if I stray, he doth convert
    And bring my minde in frame:
And all this not for my desert,
    But for his holy name.

Yea, in deaths shadie black abode
    Well may I walk, not fear:
For thou art with me; and thy rod
    To guide, thy staffe to bear.

Nay, thou dost make me sit and dine,
    Ev'n in my enemies sight:
My head with oyl, my cup with wine
    Runnes over day and night.

Surely thy sweet and wondrous love
Shall measure all my dayes;
And as it never shall remove,
So neither shall my praise.

54
The Heron

ANONYMOUS

The heron flew east, the heron flew west,
She bore her over the fair forest.
Lully, lullay; lully, lullay;
The falcon hath stolen my mate away.

She bore her up, she bore her down,
She bore her over the heath so brown.

She bore her over the meadows green,
And all to espy what might be seen.

O, then she saw an orchard fair,
Where groweth many an apple and pear.

And in that orchard there standeth a hall,
That hangèd was with purple and pall.

And in that hall there standeth a bower,
Was all clad over with lily flower.

And in that bower there standeth a bed,
With silken sheets and gold so red.

And in that bed there lieth a knight,
Whose wounds are bleeding both day and night.

And under that bed there runneth a flood,
Was half of water and half of blood.

And by that bed there standeth a stone,
And a leal maiden was set thereon.

With silver needle and silken thread,
Stemming the wounds where they did bleed.
Lully, lullay; lully, lullay;
The falcon hath stolen my mate away.

55

Autumn Change

JOHN CLARE

The leaves of autumn drop by twos and threes,
And the black cloud hung o'er the old low church
Is fixed as is a rock that never stirs.
But look again and you may well perceive
The weathercock is in another sky,
And the cloud passing leaves the blue behind.

Crimson and yellow, blotched with iron-brown,
The autumn tans and variegates the leaves;
The nuts are ripe in woods about the town;
Russet the cleared fields where the bindweed weaves
Round stubbles and still flowers; the trefoil seeds
And troubles all the lands. From rig to furrow
There's nothing left but rubbish and foul weeds.
I love to see the rabbits' snug-made burrow
Under the old hedge-bank or huge mossed oak
Claspt fast with ivy—there the rabbit breeds
Where the kite peelews and the ravens croak
And hares and rabbits at their leisure feed,
As varying autumn through her changes runs,
Season of sudden storms and brilliant suns.

56

The Little Cart

CH'ÊN TSŬ-LUNG

Translated by Arthur Waley

The little cart jolting and banging through the
yellow haze of dusk;
The man pushing behind, the woman pulling in front.
They have left the city and do not know where to go.
'Green, green, those elm-tree leaves; *they* will cure
my hunger,
If only we could find some quiet place and sup
on them together.'

The wind has flattened the yellow mother-wort;
Above it in the distance they see walls of a house.
'*There* surely must be people living who'll give you
something to eat.'
They tap at the door, but no one comes; they look in,
but the kitchen is empty.
They stand hesitating in the lonely road and their tears
fall like rain.

57

The Maidens Came

ANONYMOUS

The maidens came
When I was in my mother's bower;
I had all that I would.
The bailey beareth the bell away;
The lily, the rose, the rose I lay.

The silver is white, red is the gold;
The robes they lay in fold.
  The bailey beareth the bell away;
  The lily, the rose, the rose I lay.

And through the glass window shines the sun.
How should I love, and I so young?
  The bailey beareth the bell away;
  The lily, the rose, the rose I lay.

58

Charm

BEN JONSON

The owl is abroad, the bat, and the toad,
  And so is the cat-a-mountain;
The ant and the mole sit both in a hole,
  And the frog peeps out o' the fountain.
The dogs they do bay, and the timbrels play,
  The spindle is now a-turning;
The moon it is red, and the stars are fled,
  But all the sky is a-burning:
The ditch is made, and our nails the spade,
With pictures full, of wax and of wool;
Their livers I stick with needles quick;
There lacks but the blood to make up the flood.
Quickly, dame, then, bring your part in!
Spur, spur, upon little Martin,
Merrily, merrily, make him sail,
A worm in his mouth and a thorn in his tail,
Fire above, and fire below,
With a whip in your hand, to make him go!

59

A Tale

EDWARD THOMAS

There once the walls
Of the ruined cottage stood.
The periwinkle crawls
With flowers in its hair into the wood.

In flowerless hours
Never will the bank fail,
With everlasting flowers
On fragments of blue plates, to tell the tale.

60

There was a Man of Double Deed

ANONYMOUS

There was a man of double deed
Who sowed his garden full of seed;
And when the seed began to grow,
'Twas like a garden full of snow;
And when the snow began to fall,
Like birds it was upon the wall;
And when the birds began to fly,
'Twas like a shipwreck in the sky;
And when the sky began to crack,
'Twas like a stick upon my back;
And when my back began to smart,
'Twas like a pen-knife in my heart;
And when my heart began to bleed,
Then I was dead—and dead indeed.

61

The Sun has Long Been Set

WILLIAM WORDSWORTH

The sun has long been set,
  The stars are out by twos and threes,
The little birds are piping yet
  Among the bushes and trees;
There's a cuckoo, and one or two thrushes,
And a far-off wind that rushes,
And a sound of water that gushes,
And the cuckoo's sovereign cry
Fills all the hollow of the sky.
  Who would go parading
In London, and masquerading
On such a night of June
With that beautiful soft half-moon,
And all these innocent blisses?
On such a night as this is!

62

Last Week in October

THOMAS HARDY

The trees are undressing, and fling in many places—
On the gray road, the roof, the window-sill—
Their radiant robes and ribbons and yellow laces;
A leaf each second so is flung at will,
Here, there, another and another, still and still.

A spider's web has caught one while downcoming,
That stays there dangling when the rest pass on;
Like a suspended criminal hangs he, mumming
In golden garb, while one yet green, high yon,
Trembles, as fearing such a fate for himself anon.

63

They call me Hanging Johnny

ANONYMOUS

They call me Hanging Johnny—
Away, boys, away!
They say I hanged up many—
Then hang, boys, hang!

They say I hanged my brother—
Away, boys, away!
They say I hanged my mother—
Then hang, boys, hang!

They say I hanged my Annie—
Away, boys, away!
I hanged her up so canny—
Then hang, boys, hang!

They say I hanged my daddy—
Away, boys, away!
But I never hanged no body—
Then hang, boys, hang!

64

Upon a Spider Catching a Fly

EDWARD TAYLOR

Thou sorrow, venom Elfe:
Is this thy play,
To spin a web out of thyselfe
To Catch a Fly?
For why?

I saw a pettish wasp
  Fall foule therein:
Whom yet thy whorle pins did not hasp
  Lest he should fling
    His sting.

But as afraid, remote
  Didst stand hereat,
And with thy little finger stroke
  And gently tap
    His back.

Thus gently him didst treate
  Lest he should pet,
And in a froppish, aspish heate
  Should greatly fret
    Thy net.

Whereas the silly Fly,
  Caught by its leg,
Thou by the throate took'st hastily,
  And 'hinde the head
    Bite Dead.

This goes to pot, that not
  Nature doth call.
Strive not above what strength hath got,
  Lest in the brawle
    Thou fall.

This Fray seems thus to us:
  Hells Spider gets
His intrails spun to whip Cords thus,
  And wove to nets,
    And sets.

To tangle Adams race
  In's stratagems
To their Destructions, Spoil'd, made base
  By venom things,
    Damn'd Sins.

65

Three Little Tailors

ANONYMOUS

Three little tailors dancing in a lantern
For a bit of candle;
Candle, lantern,
For a bit of candle;
Lantern, candle,
For a bit of candle.

66

The Book-Worms

ROBERT BURNS

Through and through th' inspired leaves,
  Ye maggots, make your windings;
But O respect his lordship's taste,
  And spare the golden bindings.

67

Digging

EDWARD THOMAS

To-day I think
Only with scents,—scents dead leaves yield,
And bracken, and wild carrot's seed,
And the square mustard field;

Odours that rise
When the spade wounds the root of tree,
Rose, currant, raspberry, or goutweed,
Rhubarb or celery;

The smoke's smell, too,
Flowing from where a bonfire burns
The dead, the waste, the dangerous,
And all to sweetness turns.

It is enough
To smell, to crumble the dark earth,
While the robin sings over again
Sad songs of Autumn mirth.

68

The Golden Vanity

ANONYMOUS

'Twas all on board a ship down in a southern sea,
And she goes by the name of the 'Golden Vanity';
I'm afraid that she'll be taken by this Spanish crew,
As she sails along the Lowlands,
As she sails along the Lowlands low.

Then up speaks our saucy cabin boy, without fear or joy,
Saying, 'What will you give me, if I will her destroy?'
'I'll give you gold and silver, my daughter fine and gay,
If you'll destroy her in the Lowlands,
If you'll sink her in the Lowlands low.'

The boy filled his chest and so boldly leaped in,
The boy filled his chest and then began to swim;
He swam alongside of the bold Spanish ship,
And he sank her in the Lowlands,
And he sank her in the Lowlands low.

Some were playing cards and some were playing dice,
And some were in their hammocks sleeping very nice;
He bored two holes into her side, he let the water in,
    And he sank her in the Lowlands,
    And he sank her in the Lowlands low.

The boy then swam back unto our good ship's side,
And being much exhausted, bitterly he cried,
'Captain, take me in, for I'm going with the tide,
    And I'm sinking in the Lowlands,
    And I'm sinking in the Lowlands low.'

'I will not take you in', our captain then replied,
'I'll shoot you and I'll stab you and I'll sink you in the tide,
    And I'll sink you in the Lowlands,
    And I'll sink you in the Lowlands low.'

The boy then swam around next the larboard side,
And being more exhausted, bitterly he cried,
'Messmates, take me in, for I'm going with the tide,
    And I'm sinking in the Lowlands,
    And I'm sinking in the Lowlands low.'

They hove the boy a rope and they hoisted him on deck,
They laid him on the quarter deck, the boy here soon died;
They sewed him up in canvas sack, they hove him in the tide,
    And they buried him in the Lowlands,
    And they buried him in the Lowlands low.

69

Urry, Urry, angry Ann

ANONYMOUS

Urry, urry, angry Ann,
  Mulberry wax and tarry tan:
Ink stink stidlum stew
  Nobody out but only you.

70

The Pettichap's Nest

JOHN CLARE

Well! in my many walks I've rarely found
A place less likely for a bird to form
Its nest—close by the rut-gulled wagon-road,
And on the almost bare foot-trodden ground,
With scarce a clump of grass to keep it warm!
Where not a thistle spreads its spears abroad,
Or prickly bush, to shield it from harm's way;
And yet so snugly made, that none may spy
It out, save peradventure. You and I
Had surely passed it in our walk today,
Had chance not led us by it!—Nay, e'en now,
Had not the old bird heard us trampling by
And fluttered out, we had not seen it lie,
Brown as the roadway side. Small bits of hay
Plucked from the old propt haystack's pleachy brow,
And withered leaves, make up its outward wall,
Which from the gnarled oak-dotterel yearly fall,
And in the old hedge-bottom rot away.
Built like an oven, through a little hole,
Scarcely admitting e'en two fingers in,
Hard to discern, the birds snug entrance win.
'Tis lined with feathers warm as silken stole,
Softer than seats of down for painless ease,
And full of eggs scarce bigger even than peas!
Here's one most delicate, with spots as small
As dust and of a faint and pinky red.
We'll let them be, and safety guard them well;
For fear's rude paths around are thickly spread,
And they are left to many dangerous ways.
A green grasshopper's jump might break the shells,
Yet lowing oxen pass them morn and night,

And restless sheep around them hourly stray;
And no grass springs but hungry horses bite,
That trample past them twenty times a day.
Yet, like a miracle, in safety's lap
They still abide unhurt, and out of sight.
Stop! here's the bird—that woodman at the gap
Frightened him from the hedge: 'tis olive-green.
Well! I declare it is the pettichap!
Not bigger than the wren, and seldom seen.
I've often found her nest in chance's way,
When I in pathless woods did idly roam;
But never did I dream until today
A spot like this would be her chosen home.

71
Song

WILLIAM SHAKESPEARE

When icicles hang by the wall
  And Dick the shepherd blows his nail,
And Tom bears logs into the hall,
  And milk comes frozen home in pail;
When blood is nipt, and ways be foul,
Then nightly sings the staring owl
            Tu-whoo!
To-whit, Tu-whoo! A merry note!
While greasy Joan doth keel the pot.

When all about the wind doth blow,
  And coughing drowns the parson's saw,
And birds sit brooding in the snow,
  And Marian's nose looks red and raw;

When roasted crabs hiss in the bowl—
Then nightly sings the staring owl
Tu-whoo!
To-whit, Tu-whoo! A merry note!
While greasy Joan doth keel the pot.

72

The Good Old Man

ANONYMOUS

The music is on p. 67. The lines in italics are spoken

Where are you going, my good old man?
Where are you going, my good old man?
Where are you going, my honey, my love, my dove?
The best old soul in the world.
Going to the town.

What are you going to buy, my good old man?
New dress.

How much will it cost, my good old man?
Fifty cents.

What do you want for supper, my good old man?
Sack of potatoes.

Potatoes will kill you, my good old man?
Don't care; want to die anyhow.

Where do you want to be buried, my good old man?
In the chimney-corner.

Ashes'll fall on you, my good old man?
Don't care if they do,
So I'm close to you.

73

Summer

CHRISTINA ROSSETTI

Winter is cold-hearted,
Spring is yea and nay,
Autumn is a weather-cock
Blown every way:
Summer days for me
When every leaf is on its tree.

When Robin's not a beggar,
And Jenny Wren's a bride,
And larks hang singing, singing, singing,
Over the wheat-fields wide,
And anchored lilies ride,
And the pendulum spider
Swings from side to side.

And blue-black beetles transact business,
And gnats fly in a host,
And furry caterpillars hasten
That no time be lost,
And moths grow fat and thrive,
And ladybirds arrive.
Before green apples blush,
Before green nuts embrown,
Why one day in the country
Is worth a month in town;
Is worth a day and a year
Of the dusty, musty, lag-fast fashion.

74

Cocky Robin

ANONYMOUS

The music is on p. 68

Who killed Cocky Robin?
Who killed Cocky Robin?
Me, says the sparrow,
With my little bow and arrow,
It was I, it was I.

Who seen him die?
Me, says the fly,
With my little teenty eye.
It was I, it was I.

Who caught his blood?
Me, says the fish,
With my little silver dish.

Who made the coffin?
Me, says the crane,
With my little narrow plane.

Who made the shroud-en?
Me, says the eagle,
With my little fine needle.

Who dug his grave?
Me, says the crow,
With my little spade and hoe.

Who hauled him to it?
Me, says the lark,
With my little horse and cart.

Who let him down?
Me, says the flea,
With my little limber knee.

Who pat his grave?
Me, says the duck,
With my big splatter foot.

Who preached his funeral?
Me, says the swallow,
Just as loud as I could holloa.

75

Charms and Knots

GEORGE HERBERT

Who reade a chapter when they rise,
Shall ne're be troubled with ill eyes.

A poore man's rod, when thou dost ride,
Is both a weapon and a guide.

Who shuts his hand, hath lost his gold:
Who opens it, hath it twice told.

Who goes to bed and doth not pray,
Maketh two nights to ev'ry day.

Who by aspersions throw a stone
At th' head of others, hit their own.

Who looks on ground with humble eyes,
Findes himself there, and seeks to rise.

When th' hair is sweet through pride or lust,
The powder doth forget the dust...

In shallow waters heav'n doth show;
But who drinks on, to hell may go.

76

Stopping by Woods on a Snowy Evening

ROBERT FROST

Whose woods these are I think I know.
His house is in the village though;
He will not see me stopping here
To watch his woods fill up with snow.

My little horse must think it queer
To stop without a farmhouse near
Between the woods and frozen lake
The darkest evening of the year.

He gives his harness bells a shake
To ask if there is some mistake.
The only other sound's the sweep
Of easy wind and downy flake.

The woods are lovely, dark and deep.
But I have promises to keep,
And miles to go before I sleep,
And miles to go before I sleep.

77

'You are old, Father William'

C. L. DODGSON (LEWIS CARROLL)

'You are old, Father William,' the young man said,
  'And your hair has become very white;
And yet you incessantly stand on your head—
  Do you think, at your age, it is right?'

'In my youth,' Father William replied to his son,
  'I feared it might injure the brain;
But now that I'm perfectly sure I have none,
  Why, I do it again and again.'

'You are old,' said the youth, 'as I mentioned before,
And have grown most uncommonly fat;
Yet you turned a back-somersault in at the door—
Pray, what is the reason of that?'

'In my youth,' said the sage, as he shook his gray locks,
'I kept all my limbs very supple
By the use of this ointment—one shilling the box—
Allow me to sell you a couple?'

'You are old,' said the youth, 'and your jaws are too weak
For anything tougher than suet;
Yet you finished the goose, with the bones and the beak—
Pray how did you manage to do it?'

'In my youth,' said his father, 'I took to the law,
And argued each case with my wife;
And the muscular strength, which it gave to my jaw,
Has lasted the rest of my life.'

'You are old,' said the youth, 'one would hardly suppose
That your eye was as steady as ever;
Yet you balanced an eel on the end of your nose—
What made you so awfully clever?'

'I have answered three questions, and that is enough,'
Said his father; 'don't give yourself airs.
Do you think I can listen all day to such stuff?
Be off, or I'll kick you downstairs.'

6-2

SONGS WITH MUSIC

5
The Derby Ram

CECIL SHARP, *English Folksongs from the Southern Appalachian Mountains*, vol. II, p. 186

7

The Shad

CECIL SHARP, *op. cit.*, vol. II, p. 164

34

The Sandgate Girl's Lamentation

A. L. LLOYD, *Come All Ye Bold Miners*, p. 126

72

The Good Old Man

CECIL SHARP, *op. cit.*, vol. II, p. 338

74

Cocky Robin

CECIL SHARP, *op. cit.*, vol. II, p. 299

NOTES

1 Lu Yu, a Chinese poet, lived from A.D. 1125 to 1210. Dr Arthur Waley is the famous translator of Chinese poetry and many other Chinese and Japanese works including *The Analects of Confucius*, *The Tale of Genji* by Lady Murasaki and the Nō Plays of Japan. There are several of his beautiful translations in this anthology. Of course the poems don't rhyme in English. Charms and spells are very important to primitive people, for they believe that without them their crops, or the silkworms on which their livelihood depends, may fail.

2 Comfrey, bugloss, teasle, tansy, etc., are all wild flowers, or weeds, which you can look up in any botanical reference book. Or find some in the fields if you can, with help.

3 *Cotonea-aster* = a shrub. The 'crumb-outcaster' is the man in the house, as the birds see him.

5 A song found in many versions. This one and the tune come from a book called *English Folksongs from the Southern Appalachian Mountains*. These songs were found in this remote part of the United States of America by Cecil Sharp, early this century. This version was sung to Cecil Sharp by Mrs Margaret Jack Dodd at Beechgrove, Virginia, on 25 May 1918. The song is a lying song: one version has in the last verse the lines:

> But the man that made this song, sir,
> Was the lyin' son of a bitch.

6 Walsingham is still a shrine, in Norfolk, near Holt. But it was once much more important as a place to which pilgrims went to receive grace from Our Lady of Walsingham. I suppose this poem could be read as meaning that the love of God is 'a durable fire': but I read it as meaning that the singer's love is still constant, because he is a man, and truly in love. *Dureless* means not lasting, not enduring, of no duration.

7 A song from the Appalachian Mountains sung to Cecil Sharp by Mrs Laura V. Donald at Dewey, Virginia, U.S.A., on 10 June 1918.
A *shad* is like a large herring. You can catch them in the Severn and the Irish Shannon.

9 *Climacteric:* the period of life at which a living creature begins to decline in vital force. In human beings it is between 45 and 60, and the *grand climacteric* is at 63. People once believed a man had a climacteric every 7 years, with a grand one at 7 × 9 = 63.

Latent: hidden, dormant—that is, still inside his paws, not sticking out. *Lists:* originally the palisades of a tilting ground, here the place of combat. Knights used to 'enter the lists' when they went into a tournament.

10 *Fairly:* properly: as we might say, 'it really *is* winter.'

11 *Hauptmann Kalte:* Captain Cold, in German.

15 The compiler doesn't know what *salley* means, unless it is from *sally*, a kind of willow (sallow).

27 *Sack:* a white wine formerly imported from Spain and the Canary Islands.

31 *Spyen:* put off its mother's milk; wean.
A *keel* was a flat-bottomed vessel used on the Tyne to load up colliers, and a *keelman* was a man who worked one. From A. L. Lloyd, *Come All Ye Bold Miners.*

34 Also from A. L. Lloyd, *Come All Ye Bold Miners.* Taken from Bell's *Rhymes of the Northern Bards.*

37 This is taken from *The Ballad Book*, edited by MacEdward Leach (Harper Brothers, New York, 1955). It was once a very popular forecastle song of sailors, and of lumbermen in North America. The song was sung by Dale Porter of Maine, U.S.A. to Horace P. Beck in 1947.
Eirings: Erin's, that is, of Ireland, where Waterford and Trymore are. There was a clipper ship called 'The Flying Cloud'. But this song was printed on a broadside before the ship called by that name existed, so the sailors must have added the name of the famous ship to an existing song.
I have altered the words 'brass pounded' to 'brass pounder', but even so the sense is not very clear. Perhaps the line originally said that the ship had forty-nine six-pounder guns, or some such phrase.
An *aftermain* is presumably the sail behind the mainsail. Gallants are topgallants, the sails immediately above the topsail, and *royal galyants* the sails above the topgallants.

40 For how to play this game see David Holbrook, *Children's Games.*

41 From Leach's *Ballad Book.* This ballad celebrates the fight between the colonists under Captain John Lovewell of Dunstable, Massachusetts, against the Indians at Pigwacket, near present-day Fryeburg, Maine, U.S.A. It was first printed in 1725 and refers to an episode of 1723. Lovewell and his men were given a reward of £100 for each scalp taken; in the three months before this fight they had made £1200. One writer thought this ballad was made up by a Benjamin Franklin, uncle of the famous Benjamin Franklin.

43 People used to believe that fern-seed could make you invisible.

44 *Jungle fire:* the fire in a hobo's camp or 'jungle', usually near a railroad. The centre of their social life, and the place where they cook, and boil their clothes to kill the vermin. *Handout:* a meal given to a tramp for him to take away and eat. *Boxcar:* closed-in railway goods wagon. *Railroad bulls:* railway policemen.

50 *Coal:* charcoal.

54 *Leal:* loyal.

56 Ch'ên Tsŭ-Lung: A.D. 1608–47.

64 This poem is difficult to follow, and so I give a very rough paraphrase: You miserable, venomous creature: is this your game, to spin a web out of yourself to catch a fly? What for? (that is, is this what the spider was created by Nature to do?). I saw an irritable wasp fall foul of this net, and yet your legs (so like spikes, like the spikes in the door latch) did not stab through him, because you knew that if you did that he would throw out his sting. But, as if you were afraid, you stood a little way off and with one of your smaller legs you gently stroked and tapped his back (that is, to manipulate him into closer toils and calm him ready for the *coup de grâce*). Thus you treated him gently in case he should fly into a rage, and in a furious and snake-like fury should tear up most of your web. Whereas the heedless fly, caught by its leg, you took hastily and bit dead behind the head.

So, unless we pay attention to Nature (as the spider does) and listen to her, we shall go wrong. We should not try to do more than we have equipment or strength to do. If we do we may expect to fall in the struggle.

This episode seems to present this moral to me: Hell is a spider which spins a net of tormenting temptations out of its own bowels, weaves them in nets and traps, to entangle men in hell's stratagems. It lures them to destruction, to be spoiled, made low by venomous things, damned sins. (The poet's moral doesn't really follow from his argument: but he is excited about the spider, and this makes the poem exciting for us.)

68 From Leach's *Ballad Book*. This is a story about Sir Walter Raleigh's ship from an American seventeenth-century broadside.

71 *Crabs* are crab-apples, of course.

72 From *Folksongs of the Appalachian Mountains.* Sung to Cecil Sharp by Miss Una and Sabrina Ritchie, at Hindman School, Knox County, Kentucky, 20 September 1917. Sometimes found as a children's game.

74 From *Folksongs of the Appalachian Mountains.* Sung by Master William Agy, at Barbourville, Knox County, Kentucky, 10 May 1917. *Shrouden:* shroud, garment in which people are buried.

75 *A chapter:* a chapter of the Bible.

ACKNOWLEDGEMENTS

For permission to reproduce copyright material, acknowledgement is made to the following:

Arthur Waley and Messrs Constable & Co. Ltd ('The Pedlar of Spells', 'Lament of Hsi-Chun', 'The Little Cart' from *170 Chinese Poems*); the Literary Trustees of Walter de la Mare and the Society of Authors as their representative and Messrs Faber & Faber ('A Widow's Weeds', 'Thunder', 'Miss T.', 'Tillie'); the Trustees of the Hardy Estate and Messrs Macmillan & Co. Ltd ('Birds at Winter Nightfall', 'The House of Hospitalities', 'If it's ever Spring again', 'Last Week in October' from *Collected Poems*); The Bodley Head ('Winter Warfare' from *Collected Poems* by Edgell Rickword); Mrs Edward Thomas and Messrs Faber & Faber ('Thaw', 'Tall Nettles', 'A Tale', 'Digging'); Mrs W. B. Yeats and Messrs Macmillan ('Down by the Salley Gardens', 'The Song of the Old Mother' from *Collected Poems of W. B. Yeats*); Messrs Angus & Robertson Ltd ('Frying Pan's Theology' by A. B. Paterson); Messrs Jonathan Cape Ltd and Messrs Henry Holt & Co. Inc. ('Stopping by Woods on a Snowy Evening' from *The Complete Works of Robert Frost*); Oxford University Press ('Hailstorm in May' from *Poems of Gerard Manley Hopkins*); Miss Maud Karpeles and Oxford University Press ('The Derby Ram', 'The Shad', 'Good Old Man' from *English Folksongs from the Southern Appalachians*, Collector/Arranger Cecil J. Sharp); Novello & Co. Ltd ('Dance to your Daddy', 'Lady Maisry', 'Three Little Tailors', 'Cocky Robin', Collector/Arranger Cecil J. Sharp); Messrs Angus & Robertson Ltd ('The Crows Kept Flyin' Up' from *Old Bush Songs* collected by Douglas Stewart and Nancy Keesing); Messrs William Heinemann ('If All the World were Paper', 'The Heron' from *English and Scottish Ballads*, ed. by Robert Graves); Messrs Harper Bros ('The Flying Cloud', 'Lovewell's Flight', 'The Golden Vanity', from *The Ballad Book*, ed. by MacEdward Leach); A. L. Lloyd ('A Little Boy Threw', 'The Big Rock Candy Mountains' from *Corn on the Cob*); A. L. Lloyd and Messrs Lawrence & Wishart Ltd ('Cushie Butterfield' (rewritten by David Holbrook) and 'The Sandgate Girl's Lamentation' from *Come All Ye Bold Miners*); Curwen Edition ('Hanging Johnny'); The Folk-Lore Society ('Tarra Ding Ding Ding Dido', 'Game Rhyme' from *The Games and Diversions of Argyllshire*).

INDEX OF AUTHORS

The references are to pages